5th edition

geog.3

geog.3

ISBN 978-0-19-848992-4

9 780198 489924

geog.3 workbook 5th edition
Pack of 10

ISBN 978-0-19-848992-4

OXFORD
UNIVERSITY PRESS

workbook

<justin woolliscroft>
<katy patchwood>

Name: ....

Class: ....

OXFORD

OXFORD
UNIVERSITY PRESS

Great Clarendon Street, Oxford, OX2 6DP, United Kingdom

Oxford University Press is a department of the University of Oxford. It furthers the University's objective of excellence in research, scholarship, and education by publishing worldwide. Oxford is a registered trade mark of Oxford University Press in the UK and in certain other countries

Authors: Justin Woolliscroft, Katy Patchwood

First published in 2006

New editions published 2009 and 2015

This edition published 2023

British Library Cataloguing in Publication Data
Data available

ISBN 978-0-19-848993-1

10 9 8 7 6 5

Paper used in the production of this book is a natural, recyclable product made from wood grown in sustainable forests.

The manufacturing process conforms to the environmental regulations of the country of origin.

Printed in Great Britain by Ashford Colour Ltd, Gosport.

MIX
Paper | Supporting responsible forestry
FSC® C011748

Acknowledgements

The publisher and authors would like to thank the following for permission to use photographs and other copyright material:

Photos: p6: Mark Richardson / Alamy Stock Photo; **p10:** RichardBaker / Alamy Stock Photo; **p29:** Andrew Cole / Alamy Stock Photo; **p46(a):** PORNCHAI KITTIWONGSAKUL / AFP via Getty Images; **p46(b):** ROMEO GACAD / AFP via Getty Images; **p46(c):** A.S. Zain / Shutterstock; **p46(d):** Bullit Marquez / Associated Press / Alamy Stock Photo; **p48:** Antonio Capone / AGF / Universal Images Group via Getty Images; **p50:** Holger Mette / iStock; **p51:** Images & Stories / Alamy Stock Photo.

Artwork by Phoenix Photosetting, Q2A Media, Mike Parsons (Barking Dog Art), Ian West, Mike Phillips, Szilvia Szakall (all Beehive Illustration), Simon Tegg (Simon Tegg Illustration), Our World in Data (CC BY 4.0), Mike Connor, Steve Evans and Tracey Learoyd (Oxford University Press).

Some pages or activities in this edition are based on material written by Anna King, Catherine Hurst, and Katherine James for the original edition.

Note that the content of any direct speech attributed to characters in this book is based on information from reliable sources.

Contents

From rock to soil

1.1 Our rocky home

This is about the rocks that are beneath our feet.

1 Complete the crossword to show your knowledge of rocks. Use the information on pages 6–7 in the student book to help you.

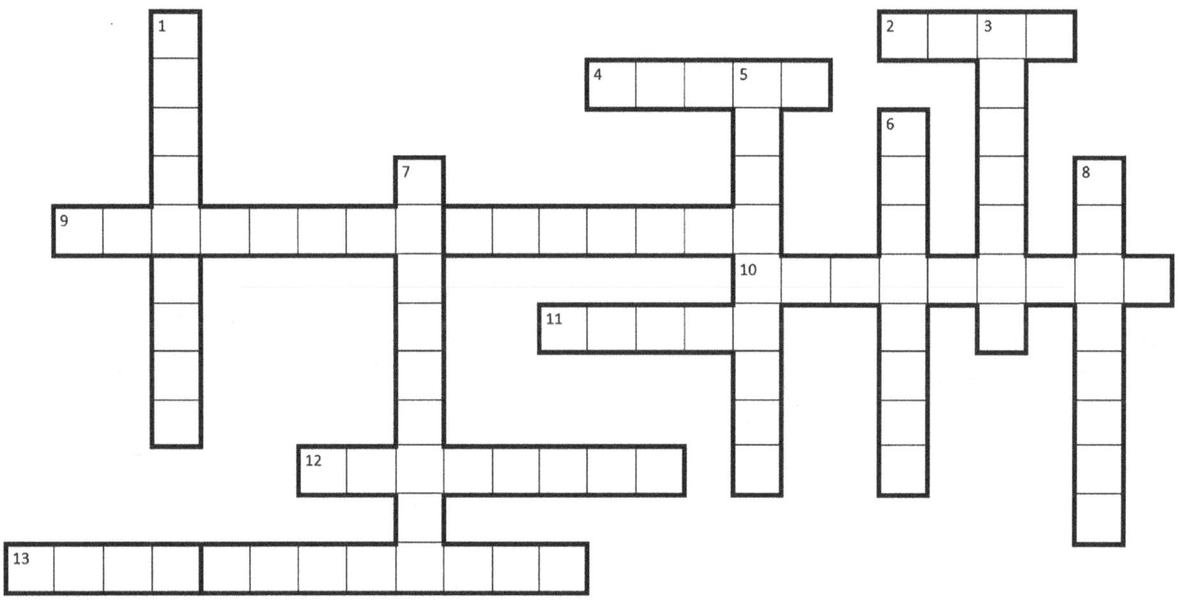

Clues

Across

2 A mixture of minerals

4 A common form of limestone

9 Another name for quartz

10 This rock is mainly quartz. The crystals are very small

11 The hard outer layer of the Earth

12 Natural compound that forms rocks

13 The number of known minerals

Down

1 Rocks from which we get metals

3 Another name for calcium carbonate

5 Rock that is mainly calcium carbonate

6 Minerals made of silicon and oxygen

7 Attractive minerals such as diamond, ruby or jade

8 Material made of crushed rock and sand

2 Go to the British Geological Survey's website and look up the Geology Viewer tool. Find out what rocks lie beneath your feet and write a description of the geology of your area below.

1.2 What are the three rock groups?

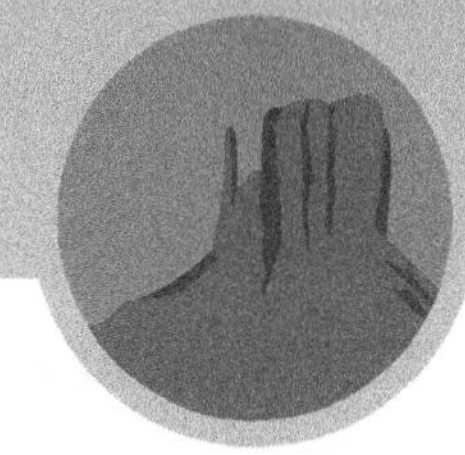

This is about igneous rocks.

1 Complete the sentences below. Use the words in the box to help you.

lock	lava	slowly	melt	crystals	magma	hot	quickly

Deep in the ground it is so .... that minerals in the rock .... giving a liquid called .... The magma then often rises and cools. It may cool .... below ground and the minerals form large .... which .... together forming rocks like granite. Some magma does reach the earth's surface, and spurts out of volcanoes as .... This then cools .... to form rocks like basalt.

2 Igneous rock forms when rock melts, then cools and hardens once again. The elements that make up all rocks are never created or destroyed; instead they are constantly being recycled.

In the box below design a logo or symbol that could be used tell people this fact about rocks – the oldest recycling machine on the planet! Be as creative and inventive as you can!

1.3 What is weathering?

This is about weathering and how it shapes rocks.

1 Lions are the kings of the jungle because of their raw power and strength; however, nothing can stop the weather!

Look at the picture of the stone lion which is found in the city of Nottingham. Over many years, the lion will slowly wear away because of weathering. Describe how different the lion might look in 500 years. What would the lion think about it?!

2 Write your own definition of the three main types of weathering. Use your definitions to play charades with somebody else in your class. Can they guess which type of weathering you are acting out?

Physical weathering ....

Chemical weathering ....

Biological weathering ....

1.4 What is the rock cycle?

Here you will learn more about why 'recycling rocks'!

1 Draw a line to match each word with its meaning.

Word	*Meaning*
Magma	liquid rock at or above the Earth's surface
Erosion	particles of eroded rock or plant and animal debris
Sediment	liquid rock below the Earth's surface
Lava	the process of wearing away rocks

2 Are these statements true or false? Put a tick in the correct box.

		True	False
a	The hard outer part of Earth is cracked into huge slabs, called plates	☐	☐
b	Some slabs collide because of plate movements	☐	☐
c	The rock cycle destroys materials such as granite	☐	☐
d	Diamonds form when magma cools in sedimentary rock	☐	☐
e	Weathering leads to soil	☐	☐

3 Describe in your own words what the diagram below shows.

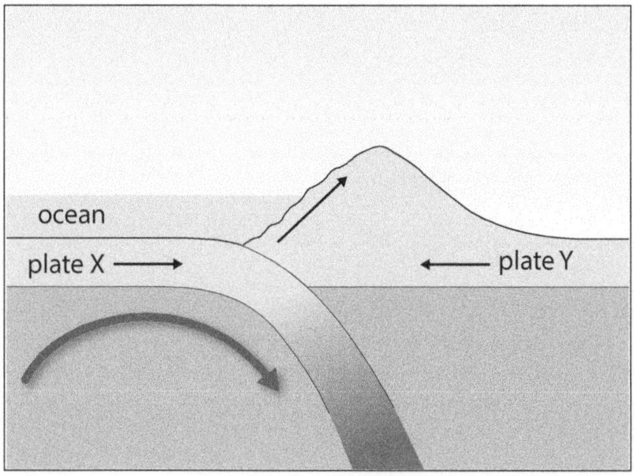

....

....

....

....

1.5 The British Isles on their travels!

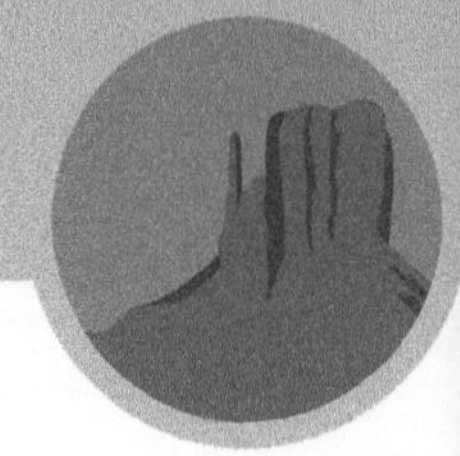

Here you will explore the geological timescale.

1 Printed below are nine events that have happened in geological history. Put them in the correct order by writing a number from 1–9 at the start of each box.

Dinosaurs become extinct	Mammals and birds thrive	First animals with shells appear
First living cells appear in sea	Giant insects emerge	First birds appear
Warm-blooded reptiles appear	Dinosaurs first appear	Humans appear and spread

2 Using the words in the boxes and information from page 15 in **geog.3** to help you, write a paragraph describing how the earth evolved. Remember to use years and the correct names of the geological periods in your answer.

1.6 Rock and relief around the UK

Here you will find out some more about the mountains in the UK.

1 Read the following descriptions of mountains in the UK. Use an atlas and do your research to find out the name of each of the mountains, and their relative height (their height in relation to their surrounding). Write your answers in the table below.

Who am I?	**Mountain name**	**Relative height in metres**
I am on the west coast of Scotland at the western end of the Grampian mountains. I am close to the town of Fort William, and am proud to be the tallest mountain in the British Isles.		
I am the highest mountain in northern Scotland, north of the Great Glen. I am found west-south-west of Inverness. I am very isolated, lying 10km from the nearest road.		
I can be found near the town of Porthmadoc in north Wales. There is a national park named after me. I am the highest point in the British Isles outside of Scotland. I have a railway too!		
I am found right in the middle of the Cairngorms National Park and am the second highest mountain in the United Kingdom. They say that I am haunted by a big Grey Man, but I'm not telling!		
Hello! You will find me in the English county of Cumbria and I am the highest mountain in England! Climb me on a clear day, and you can see Scotland, Wales, Ireland, and the Isle of Man! I am found to the east of Wastwater, one of the famous lakes in the Lake District National Park.		
I am found between Thirlmere and Ullswater, two lakes in the Lake District National Park. I lie north-east of the highest mountain in England and south-east of Keswick.		

2 On the map below the six mountains are numbered 1–6. Write the name of each mountain in the space next to the number.

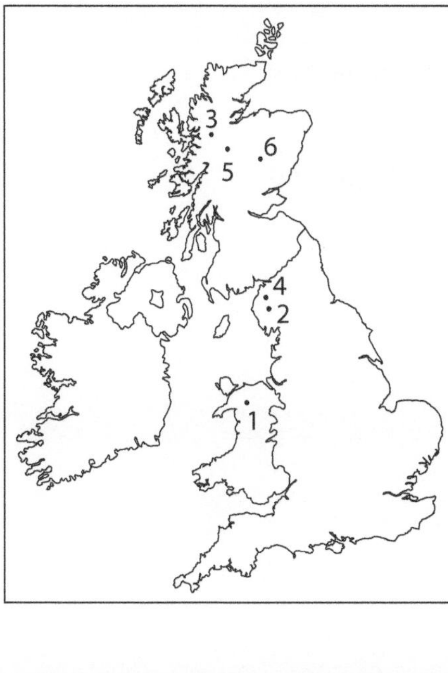

Number	**Mountain name**
1	
2	
3	
4	
5	
6	

1.7 What's the link between rock and landscape?

Here you will find out some more about limestone landscapes.

1 How does the acid found in rain affect the limestone?

2 Why is the soil thin in limestone areas?

Tip!

As the limestone dissolves, it often leaves gaps where the joints in the limestone are. It can form a limestone pavement as shown in the photograph. The gaps are called *grykes* and the blocks are called *clints*.

3 Annotate the photograph above to show where the grykes and clints are found. For each, explain how they are formed.

4 Many limestone areas have underground caves. Why do you think caves form underground?

1.8 How are *you* linked to soil?

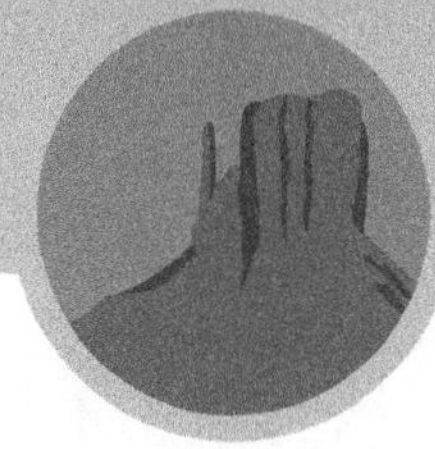

Here you will find out why worms are so important.

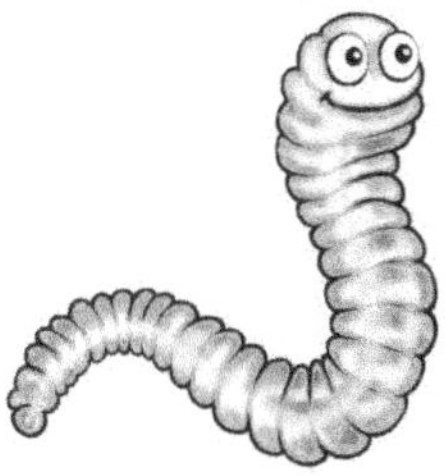

Hello, I'm Willow the worm!

I have such a busy life, even though I move quite slowly. I have even been called an 'ecosystem engineer'! I can make burrows that are vertical or go sideways, allowing all that lovely oxygen and water to come into the soil.

I don't have a problem choosing what to eat, like you humans do – if it is rotting, decaying, or organic I will want to have a taste! A lot of my day is spent 'decomposing' – eating and breaking up organic matter such as dead plants and animals so that bacteria and fungi can feed on it and release the lovely nutrients back into the soil. With all this eating it is not surprising that I leave behind castings that are a very valuable type of fertiliser. And that's not all, because as I move I also mix up the soil layers. A famous scientist called Charles Darwin called us 'nature's ploughs' and others have called us 'tiny food processors'!

1 Why do you think worms have been called 'ecosystem engineers'?

....

....

2 How can worm burrows help plants grow?

....

....

3 Where do the nutrients come from that go back into the soil after the worm spends its time decomposing?

....

....

4 Why do you think Darwin called worms 'nature's ploughs'? Do you think that's a good name for them? Why?

....

....

5 In the space below draw a poster that tells people why worms are so important. Be as creative as you can!

Using Earth's resources

2.1 Earth's natural resources

Here we look at our basic needs and the resources used to cover them.

This Earth we live on provides us with our basic needs (see page 24 in **geog.3**):

Water Food Clothing Shelter Energy

1 Write down how your own basic needs are met.

Water ....

....

Food ....

....

Clothing ....

....

Shelter ....

....

Energy ....

....

2 Can you define the following terms and give an example of each?

a Natural resource: ....

Example: ....

b Renewable resource: ....

Example: ....

c Non-renewable resource: ....

Example: ....

3 The world's population keeps rising. Why does this pose a challenge when it comes to resources?

....

....

Water part 1 – Do we have enough?

This is about how increasing population and wealth affect water use.

1 Over the next 30 years, the global demand for water is expected to increase.

Water use	2030 (% change)	2050 (% change)
Farming	−3.6	−6.9
Industry	+53.3	+119.8
Domestic Use	+36.5	+65.1

Add the three missing bars to the graph, using data from the table.

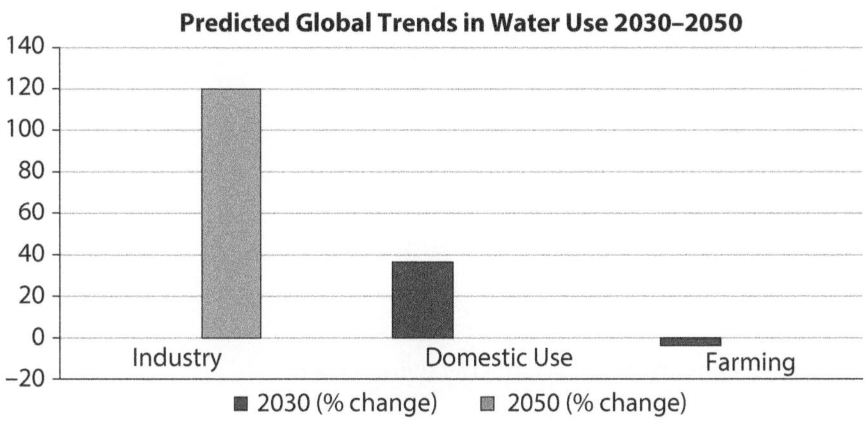

2 Earth's population is increasing rapidly. Explain how this may affect:

Industrial water use ....

....

Domestic water use ....

....

3 People are becoming richer, and the middle class is expanding. Explain how this may affect:

Industrial water use ....

....

Domestic water use ....

....

4 The data predicts that water use for farming will decrease. What could be the reason for this?

....

....

2.3 Water part 2 – Emptying the Ogallala?

Here we look at the future for farmers when their underground irrigation water runs out.

1 What ideas do you think Brad might come up with to ensure that he can stay living where he is? Check your ideas against suggestions in unit 2.4.

2 Find out where the largest groundwater aquifer in the world is. What makes it less likely to run out of water in the near future compared to the Ogallala?

2.4 Water part 3 – Tackling water stress

Here we look at how the problem of economic water scarcity is being addressed.

WaterAid

WaterAid is an international organisation whose mission is to transform the lives of the poorest and most marginalised people by improving access to safe water, sanitation, and hygiene.

1 Find out about WaterAid. Write down in your own words what its aims are.

2 Make a list of the different ways in which WaterAid has set about achieving its mission.

3 Think about your day to day life. How much water do you use in a day? What could you change to reduce your water consumption?

4 Define the following terms:

a harvesting pond

b drip irrigation

c desalination

2.5 Food part 1 – Feeding the UK

Here we look at food security in the UK.

1 Overall, around 64% of the food we eat in the UK is produced here, while the rest is imported. Look at the list of food items below, and decide whether the UK mostly produces them or imports them. Then, draw each food item in the correct box.

wheat	banana	egg	tea	orange	lemon	potato

Food largely produced in the UK	Food largely imported by the UK

2 Write one advantage and one disadvantage of food imports.

Advantage: ....

Disadvantage: ....

3 What impact could climate change have on food security in the UK? List at least two possible consequences.

....

....

....

....

4 One of the changes the UK could make to become more self-sufficient in food is use hydroponics. Explain what that consists of.

....

....

2.6 Food part 2 – Food insecurity, a global issue

This is about food insecurity, and the issues that lead to it.

1 Decide whether these statements are true or false, then put a tick in the correct box.

		True	False
a	Some countries with high levels of food insecurity have lots of farmers.	☐	☐
b	Famine and food insecurity are the same thing.	☐	☐
c	The only factors linked to food insecurity are poverty levels.	☐	☐
d	Poor infrastructure can make it harder to distribute food.	☐	☐

2 If you marked any of the statements from exercise 1 as false, explain why they are incorrect.

....

....

....

....

3 Look at this world map, which shows hunger (a strong indicator of food insecurity) around the world.

Hunger around the world

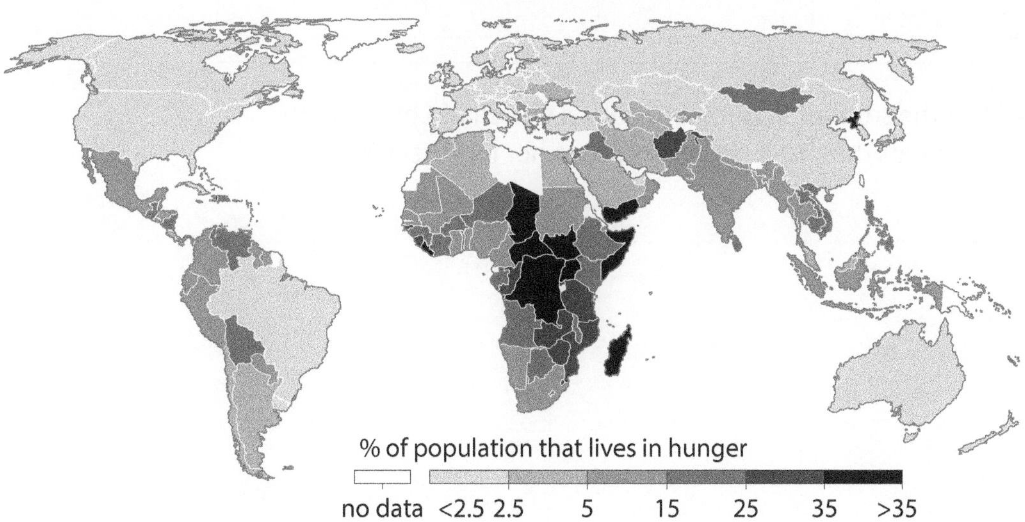

As you can see, hunger is not distributed evenly around the world. In the countries coloured in a darker shade, many people suffer from food insecurity. Why do you think that is? List at least three factors that contribute to food insecurity there.

....

....

....

....

2.7 Food part 3 – Tackling food insecurity

This is about ways to achieve food security in developing countries.

1 Food security is when people have access, at all times, to the food they need.

In each of the boxes below, write a sentence explaining how people in developing countries can get more food security by using farming methods, technology, and storing food.

Use at least two of these words and phrases in each sentence. Tick them off when you have used them:

☐ sacks	☐ artificial intelligence	☐ drones	☐ insects
☐ fertiliser	☐ satellites	☐ seeds	☐ rainwater pits

Farming methods

....

....

....

Technology

....

....

....

....

....

....

....

Food storage

....

....

....

2 Read about Chifundo on page 36 of the student book, then make a sketch of Chifundo on her farm in Malawi. Include all the things that have helped improve her food security. Label each feature.

2.8 Energy part 1 – Goodbye to the fossil fuels?

This looks at how future generations will judge our reliance on oil.

1 As our use of oil goes up, there are fewer supplies of it left in the ground. Can you think of two reasons for leaving it in the ground for future generations?

2 How do you think students of your age in a hundred years' time will look back on our use of oil?

3 You are a writer in the year 2115. By then all our energy will come from renewable sources. Write an article below titled 'The fossil fuel age'.

2.9 Energy part 2 – Renewables in the UK

This is about the natural sources of energy in the UK.

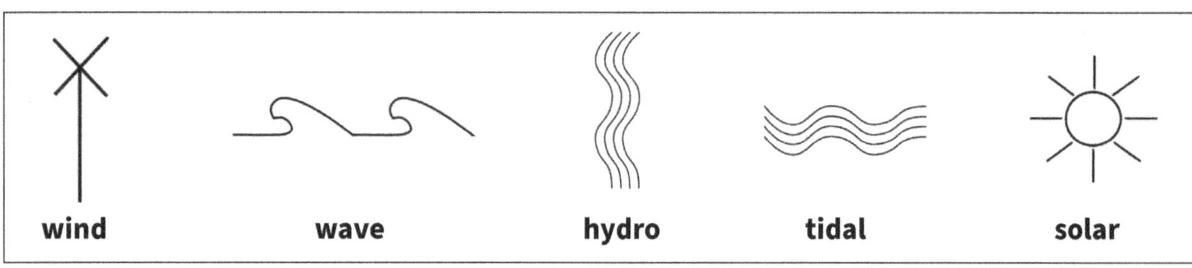

1 Imagine it is 2050. How would this town get its energy? Draw symbols on the town showing what you think should happen.

2 The town has decided to get all of its energy from renewable sources. What problems might there be?

....

....

....

....

....

2.10 Energy part 3 – Is solar power the solution?

This explains what solar power is, and where it can best be used.

1 Circle the correct word below to complete the sentence.

Solar power means using energy from the SEA / RAIN / SUN / WIND.

2 Use the map to give the places in the table a ranking for solar power. Give number 1 to the place best suited for solar power, down to 6 for the place least suited.

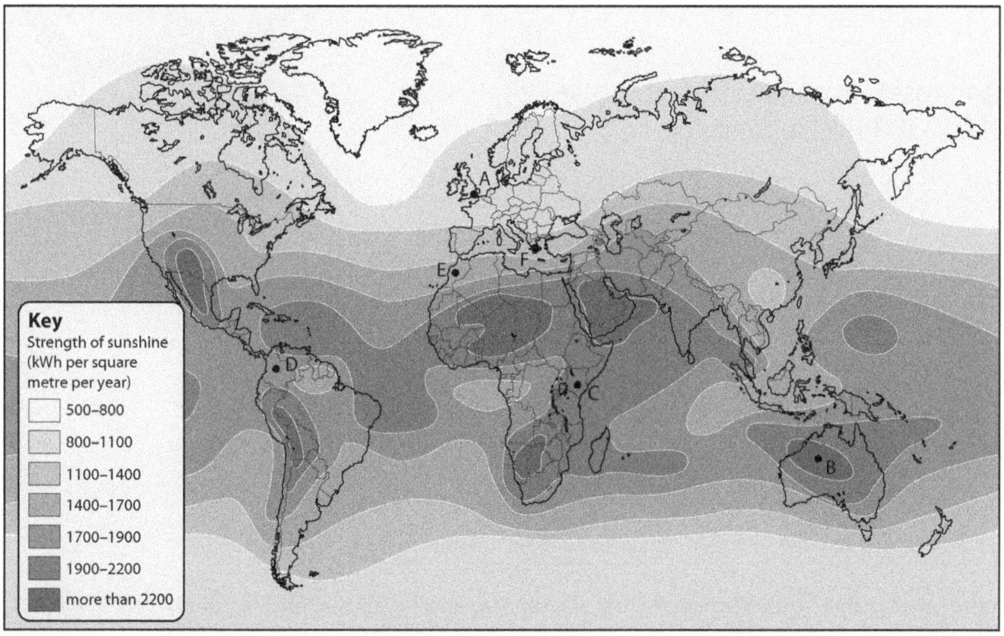

Location on map	Name	Ranking
A	Britain	
B	Australia	
C	Kenya	
D	Colombia	
E	Morocco	
F	Greece	

Solar power is useful because the PV cells can convert sunlight into electrical energy on a stand-alone system. This means they don't need to be connected to a national grid and can be sited anywhere (providing there is enough sunlight!).

3 Complete the two speech bubbles for Jomo, a Kenyan schoolboy living in an isolated small village. What does he think about the possibilities of solar power?

My country is sunny all year round, so ...

....

....

....

....

My school is miles from anywhere, but ...

....

....

....

....

Why must we change our ways?

Here we consider the threats from human activity to biodiversity.

The word that is used to express the number and variety of animal and plant species is 'biodiversity'. One term that is used to express the greatest threats to biodiversity is **HIPPO**, standing for **H**abitat destruction, **I**nvasive species, **P**ollution, human over-**P**opulation, and **O**ver-harvesting.

1 Refer to pages 44–45 in **geog.3** and research on the internet to give examples of each of the categories that HIPPO represents.

Habitat destruction ....

....

....

....

Invasive species ....

....

....

....

Pollution ....

....

....

....

human over-**P**opulation ....

....

....

....

Over-harvesting ....

....

....

....

Earning a living

3.1 What kind of work do we do, in the UK?

This is about the kind of work people in the UK do for a living.

1 What is an *employment sector*?

....

....

2 a Circle the correct word to complete each sentence.

In the SECONDARY / PRIMARY / QUATERNARY sector, people gather materials from the earth.

In the SECONDARY / QUATERNARY / TERTIARY sector, people do hi-tech research.

In the PRIMARY / TERTIARY / QUATERNARY sector, people provide a service for others.

In the TERTIARY / SECONDARY / PRIMARY sector, people turn materials into things to sell.

b Now give two example of jobs in each sector.

Primary:

....

Secondary:

....

Tertiary:

....

Quaternary:

....

3 Imagine that there are big strikes in the forestry industry, in the primary sector. Do you think that would have an impact on the secondary sector? Explain why.

....

....

....

....

....

3.2 The UK's changing pattern of employment

Here we consider what changes in our employment structure lie ahead.

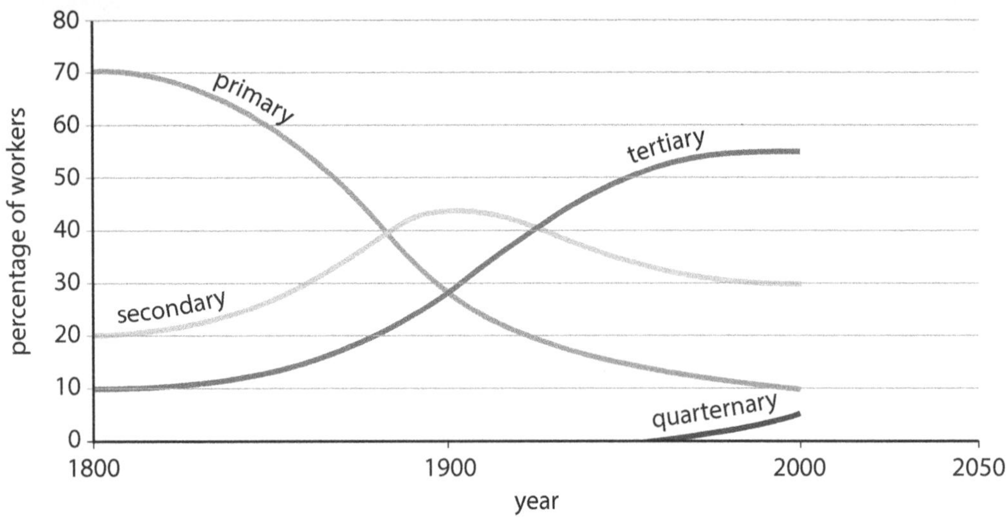

Look at the graph, which appeared in **geog.3**. Extend the lines according to how you think the graph might look in 2050. Consider the following:

- Our farmers still produce 60% of the food we eat.
- Manufacturing has started to return to the UK.
- Automation in the service sector is increasing.

Don't forget that the four lines have to add up to 100%!

1 Although our agricultural sector is still an important part of the economy, it represents a very small percentage of the workforce. Why is this?

....

....

2 Factory workers in less-developed countries are becoming more prosperous. How might this affect manufacturing in the UK?

....

....

3 If manufacturing begins to return to the UK, will it provide the same number of jobs as before? If not, why not?

....

....

4 The UK leads the world in the creation and design of computer games. Why do you think that this activity can be described as part of the quaternary sector?

....

....

3.3 Example: change in and around Doncaster

Here we look at the different sorts of jobs that Doncaster can offer.

1 All three of the people above live in Doncaster. Complete their statements.

2 Which of the three is the most likely to leave Doncaster in search of work? Why?

....

....

3 Research the tourist attractions in and around Doncaster. Given its long history and the many local attractions, what do you think could be done to boost Doncaster's tourist industry?

....

....

....

....

....

3.4 Why do employment patterns change?

This is about the changes in employment patterns and the role that technology played in those.

1 Draw lines to link each of these big drivers of change with its correct definition.

Technology	Where we put scientific discoveries and inventions to practical use.
Globalisation	The people that run a country can offer finance for factories, grants for research, grants for development, etc.
Governments	The increasing movement of goods, services, people money, and ideas, around the world.

2 Give an example of technology that each of the following workers might use. Then, think about which sector each of them works in, and put a tick in the correct box. The first one is done for you.

	Primary sector	Secondary sector	Tertiary sector	Quaternary sector
a Scientist: *a new powerful microscope*	☐	☐	☐	✓
b Builder:	☐	☐	☐	☐
c Farmer:	☐	☐	☐	☐
d Teacher:	☐	☐	☐	☐

3 Look at the boxes you've ticked in the previous exercise. Do you think all sectors have seen some change brought on by technology, or are some of them not affected by it?

....

4 Do you think that technology has made the percentage of workers in the following sectors go up, down, or has had no effect on it? Draw the correct symbol next to each sector:

The % has gone up: ↑ The % has gone down: ↓ Technology hasn't had an effect: **X**

a Primary sector:

b Secondary sector:

c Tertiary sector:

3.5 A closer look at globalisation and the UK

Here we look at the causes and consequences of globalisation.

1 Read the following comments and decide if each person considers globalisation a benefit or a problem. Then, draw a happy face ☺ or a sad face ☹ accordingly in each circle.

◯ A lovely local business that had been in my town for ages had to close down because it couldn't compete with TNCs. I miss it!

I lost my job in IT because it was cheaper for the company to outsource the work than to pay UK wages. ◯

◯ I work in the hospitality sector and most of my clients are tourists. They bring a lot of money into the country!

We have a wider choice when shopping, and the prices for things made abroad are often lower than for things made in the UK. ◯

2 Containerisation, information technology, and air travel all played a part in globalisation. Pick one, and explain how it played a part.

....

....

....

3 Governments can favour globalisation if it helps their economies, or they can slow it down to protect their industries. Read the following measures and decide it they are in place to push globalisation, or to slow it down.

	Push	**Slow down**
a impose tariffs	☐	☐
b make trade deals	☐	☐
c refuse work permits to people from overseas	☐	☐
d limit access to social media	☐	☐

4 Can you list some of the ways in which the Covid-19 pandemic slowed down globalisation?

....

....

....

3.6 Is your mobile linked to globalisation?

This is about TNCs and the arguments for and against globalisation.

1 Which employment sectors are involved in transnational corporations (TNCs), such as Apple? Tick all that apply.

The primary sector	☐	The tertiary sector	☐
The secondary sector	☐	The quaternary sector	☐

2 a Look at the table. Decide which arguments are For and which Against globalisation. Circle your answer.

b Then decide which arguments can be described as Social, Economic, or Environmental. You may want to circle more than one possible answer.

Argument	For or against?	Social (S), economic (Ec), and/or environmental (En)?
Globalisation breaks down barriers between countries.	For / Against	S / Ec / En
Globalsation causes unemployment in some developed countries.	For / Against	S / Ec / En
When a multinational opens a factory in a developing country, it helps that country to develop.	For / Against	S / Ec / En
The activities of some multinationals can pollute an area.	For / Against	S / Ec / En
Multinationals look for the cheapest source of labour.	For / Against	S / Ec / En
Globalisation can change the culture of a developing country.	For / Against	S / Ec / En
The wages earned by people in developing countries help their country to develop.	For / Against	S / Ec / En
Many developing countries do not have adequate laws to make sure the water and air are kept clean.	For / Against	S / Ec / En
Globalisation gives people a wide choice over what they can buy.	For / Against	S / Ec / En
Multinationals help developing countries to exploit their resources.	For / Against	S / Ec / En

3 Select one argument from the table that is for globalisation, and one that is against it. For each one, write a short paragraph to support that argument.

For: ....

....

....

....

Against: ....

....

....

....

Employment and the shock of Covid-19

This is about changes to working habits during and after the Covid-19 pandemic.

1 During the lockdowns of the Covid-19 outbreak, many adults, students, and children had to do their work at home. Sometimes this was challenging, but there were also some benefits. How was it for you?

Fill in the boxes below. You can write about things such as: family, friends, meals, transport, daily routine, distractions, technology...

Things I found challenging about studying at home…	Benefits of studying at home….

2 Read what Jed has to say about his lifestyle. Fill the speech bubble.

After the pandemic, my boss said I didn't need to go back to the office, so I decided to pack up my laptop and travel around the world for a few months. I can work anywhere with a good Wi-Fi connection. I'm currently in South America. Next stop, Asia! People say I'm a 'digital nomad'.

The best things about this lifestyle are…

....

....

But sometimes, it's hard because…

....

....

3 Due to the rise in remote working, some office space is now vacant. Look at this empty office building. Try to come up with five things that the building could be used for now.

....

....

....

....

3.8 So what kind of work will *you* do?

This is about green jobs and climate change.

1 Draw lines to match up this list of green jobs with their job descriptions.

a Air quality engineer	**1** Solves issues such as hazardous waste disposal, soil quality issues, and habitat protection.	P
b Chief sustainability officer	**2** Researches improvements for farming techniques and developing new food products.	
c Environmental consultant	**3** Monitors the emissions of pollutants in the air, especially in urban areas.	
d Hydrologist	**4** Develops sustainability programmes for businesses and oversees them.	
e Food scientist	**5** Designs techniques for removing carbon from the atmosphere, including capture of CO_2 from the air and nature-based methods involving forests and microbes.	
f Wind turbine technician	**6** Solves complex problems related to water access and quality.	
g Carbon removal engineer	**7** Uses electrical and computer skills to maintain wind turbines.	

2 Green jobs have different functions, such as:

- Protecting the environment (**P**). *E.g. Preserving natural resources such as soil, oceans, rivers, air, forests.*
- Limiting climate change (**L**). *E.g. Making global agreements to restrict the use of fossil fuels.*
- Reducing the impacts of climate change (**R**). *E.g. People may grow different crops as the climate gets warmer, cooler, wetter or drier.*

Try to explain the difference between 'limiting climate change' and 'reducing the impacts of climate change.'

....

....

3 Look at the jobs from exercise 1. What are their functions? Write the letters **P**, **L**, and **R** in the empty boxes to indicate the job function. You can use more than one letter per job; the first one has been done as an example.

International development

4.1 An unequal world

This is about how unequal our world is.

1 The shaded areas in the pie graphs below show the following:

- **a** the 77% of people around the world who live on less than $10 a day;
- **b** the 24% of people who have no electricity;
- **c** the 8% of people who never get enough to eat, ever – who are starving;
- **d** the 1 in 5 people aged over 15 who can't read or write.

Match the numbers against the pies to the letters above.

1

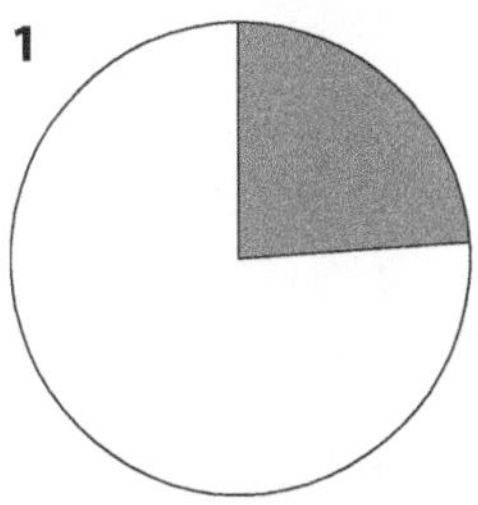

1 = ...

2

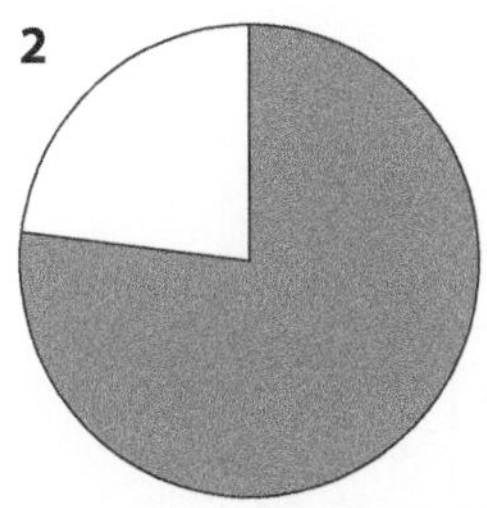

2 = ...

3

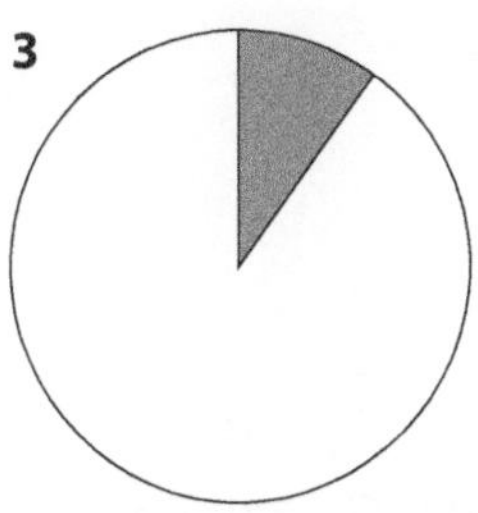

3 = ...

4

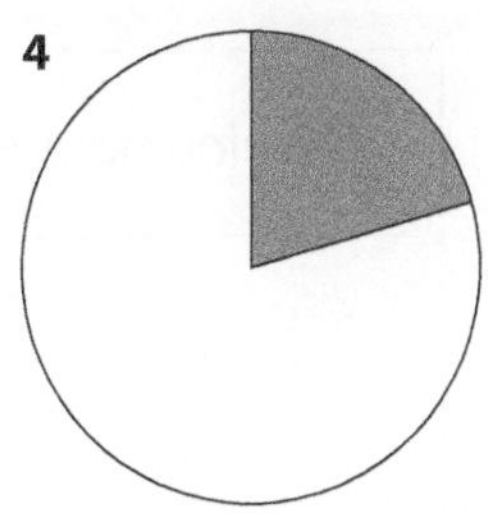

4 = ...

2 Decide if each of the following situations describes life in poverty, then tick the correct box.

	poverty	**not poverty**
a living in a home that doesn't have clean safe water	☐	☐
b having resources worth around $1.90 a day to live on	☐	☐
c sharing a toilet with another person from your household	☐	☐
d having savings for potential emergencies	☐	☐
e living in a home that has electricity	☐	☐

4.2 So... what is development?

This is about different aspects of development.

1 Complete the spider diagram below, using the words and statements from the box. (You will need to add more 'legs' to the diagram.)

☺ *education for everyone*	☺ more healthcare for everyone	☺ justice
☺ lower pollution	☺ living longer	☺ more money for the country
☺ equality for women	☺ sustainability	☺ fewer children dying
☺ more jobs		

2 Then add some more ideas of your own.

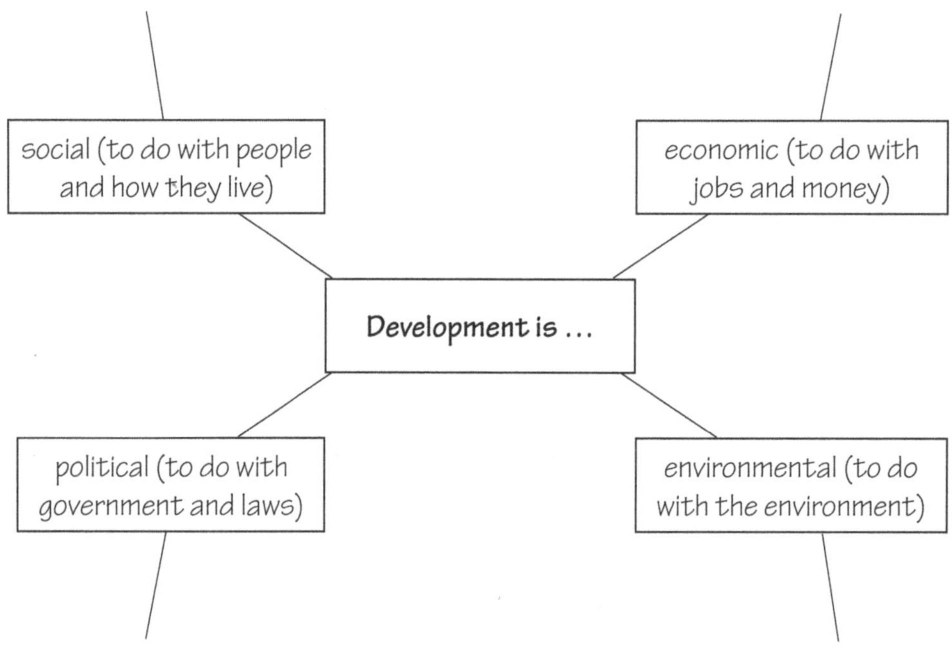

4.3 How is development measured?

This is about identifying developed and developing countries around the world.

1 Finish off these sentences. Use the phrase 'standard of living' in your sentences.

Richer countries are called developed. That means … ....

Poorer ones are called developing. That means … ....

2 If you landed in a strange country and didn't know where you were, how could you tell if it was developed? Think of at least five ways.

....

....

....

....

....

3 Look at the countries listed below. Some of them are developed, and some are developing.

You need two colours. Use one colour to colour in the developed countries, and the other one to colour in the developing countries. Cross them off the list as you go.

Australia Brazil Canada Egypt Ghana India Japan Kenya New Zealand Peru Somalia Sweden United Kingdom USA Zimbabwe

4.4 Malawi: a developing country

Here we look at Sephora's life, and at some of the problems that Malawi faces.

1 Imagine you are Sephora. In your own words, describe what you do during the day.

At 6.15 am I... ....

....

At 7 am I... ....

....

At 1 pm I... ....

....

At 6 pm I... ....

....

At 7 pm I... ....

....

2 Explain what your hopes are for the future. Are they likely to happen? If not, why not?

....

....

....

3 Malawi depends more and more on tobacco sales for export earnings. Why might this be a problem in the future?

....

....

....

4 Do the majority of workers in Malawi earn a living by working on the primary, the secondary, or the tertiary sector? What type of job do they do?

....

5 Use the following numbers to complete the text: 4, 64, 18, 51.

Around ....% of the people in Malawi live in poverty. The median age of the population is .... and the life expectancy is .... They have an average of .... doctors per 100 000 people.

4.5 How did the development gap grow? – part 1

This is about the reasons for the development gap.

1 The paragraphs below are all about why some countries are poorer than others. The words in brackets are all jumbled up though! You need to unscramble them and write them on the lines.

....

Many poorer countries were once (cilnoodes) .... by European countries. The Europeans took raw (atameirls) .... and turned them into finished (gsdoo) .... They sold these goods back to the poorer country! When the colonisers left, the countries had little (iustrnyd) ...., people didn't have much (edcaution) .... or many (ssklli) ....

....

Some poorer countries have difficult (matescli) .... so they can't grow (fdoo) ....

....

A country has a better chance of developing if it is (ecuesre) ...., but some poorer countries have (sraw) .... They also have (ropctur) .... leaders who make themselves rich while their people live in (rvoypet) ....

2 The reasons above can be divided into three groups. Match the titles of the groups with their meanings by drawing linking lines.

historical reasons	reasons that are about natural resources, rainfall, temperature, soils and landscapes
social and political reasons	reasons that are about what happened in the past, like colonisation
geographical reasons	reasons that are about politics and money

3 In the space above each paragraph in the box, you are going to write its title. Use the group titles from question **2**.

How did the development gap grow? – part 2

This is about the factors that keep the development gap wide.

Here is a list of the Heavily Indebted Poor Countries (HIPCs) that have qualified for and received financial help.

[Note. To qualify for HIPC status a country has to prove that it cannot pay off its debt even after it has received all normal debt assistance.]

Benin
Guinea
Guinea- Bissau
Honduras
Madagascar
Malawi
Mali
Mauritania
Mozambique
Nicaragua
Bolivia
Burkina Faso
Cameroon
Chad
Democratic Republic of Congo
Ethiopia
the Gambia
Ghana
Niger
Rwanda
São Tomé and Príncipe
Sierra Leone
Senegal
Tanzania
Uganda
Zambia
Afghanistan
Burundi
Centra African Republic
Comoros
Republic of Congo
Côte d'Ivoire
Guyana
Haiti
Liberia
Togo

Colour in the HIPC countries in Africa on the map above. Then cross them off the list above.

1 Which countries on the list are **not** in Africa?

....

2 Can you list at least three things that make it difficult for poor countries to escape poverty?

....

....

....

4.7 Globalisation, development and TNCs

This is about TNCs and their relationship with developing countries.

1 Complete the sentences below. Use the words in the box to help you.

profit	goods	cheaper	Chinese	headquarters	lower	(TNCS)	international

Transnational Corporations .... are large companies with .... branches. Most have their .... in high income countries, but their manufacturing is usually done in developing countries. This is because wages are .... and raw materials are .... there, so it costs less to produce ...., and they can make more .... The world's richest TNCs are American and ....

2 TNCs are responsible for over half of all world trade, and they can have a huge impact on development. But, what does development mean?

Complete the definition:

Development is... ....

3 To find out how a country is changing, we can study its **development indicators** over time.

a Life expectancy is a useful indicator. Write a definition and then explain how TNCs can have an impact.

Development indicator: Life expectancy

Definition: ....

TNCs can impact life expectancy because... ....

....

....

b Now choose another development indicator. This could be GNI, literacy rate, number of doctors per person – or something else.

Development indicator: ....

Definition: ....

TNCs can impact this because... ....

....

....

Covid-19 and vaccine inequality

This is about the Covid-19 vaccine.

1 Why can we say that creating a vaccine against Covid-19 was a global effort? List two reasons.

2 This graph shows the total of Covid-19 vaccines administered per 100 people in high-income countries, low-income countries, and in the UK. (Notice that over 200 hundred vaccine doses got administered per 100 people because some people got second doses and boosters).

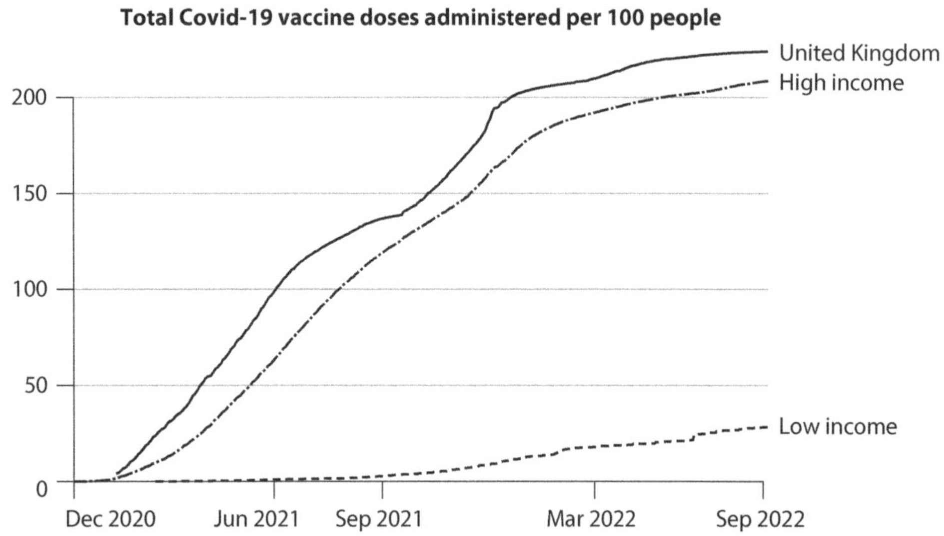

a The graph shows a big difference in the rate at which high-income countries, such as the UK, and low-income countries, vaccinated their populations. Why do you think that is?

b Looking at the graph, you can see that low-income countries slowly started to get access to more doses of the vaccine. What is the scheme that allowed for that to happen?

c As you can see in the graph, the UK was one of the first countries to administer Covid-19 vaccines. In fact, to make that happen, the UK ordered about three times what it needed. Can you list a positive and a negative result of that decision?

Escaping from poverty

This is about the different kinds of migrants.

1 Some of the people shown could be called refugees, and some migrants. Is it possible to make a distinction between them? What are your views?

4.10 Can we put an end to poverty?

Trade or aid? This looks at the arguments.

1 The arguments in favour of aid or trade to help developing countries has been going on for a long time. Read the arguments for and against below, and then give your opinions.

Arguments in favour of trade

- **Trade** makes a developing country more independent of the aid-giving countries. Aid budgets can always be cut.
- **Trade** helps developing countries to maintain their dignity. Aid can be seen as a form of charity.
- **Trade** establishes a strong impression in the international market.
- **Trade** improves the economic performance of a country while aid money can be misused. Aid can lead to corruption.

Arguments in favour of aid

- **Aid** can go directly to where it is needed, while trade can distribute resources inefficiently. The benefit of trade is mostly confined to an elite group of people within the country.
- **Aid** is not always provided in the form of money, and is sometimes provided through expert advisors. On the other hand, trade needs a good infrastructure of the country to prosper. It is very difficult for the developing countries to maintain a good infrastructure.
- **Aid** allows for money in a given country to be allocated well against need.
- Exposing fragile developing economies to free **trade** is very risky.
- **Trade** requires investment first.
- **Trade** does not necessarily mean **fair trade.**

Aid or trade?

Our restless planet

5.1 What's Earth like inside?

This is about the three layers that make up the Earth.

1 This diagram shows the layers that make up the Earth.

a Write the words from the box in the rectangles to correctly label the diagram.

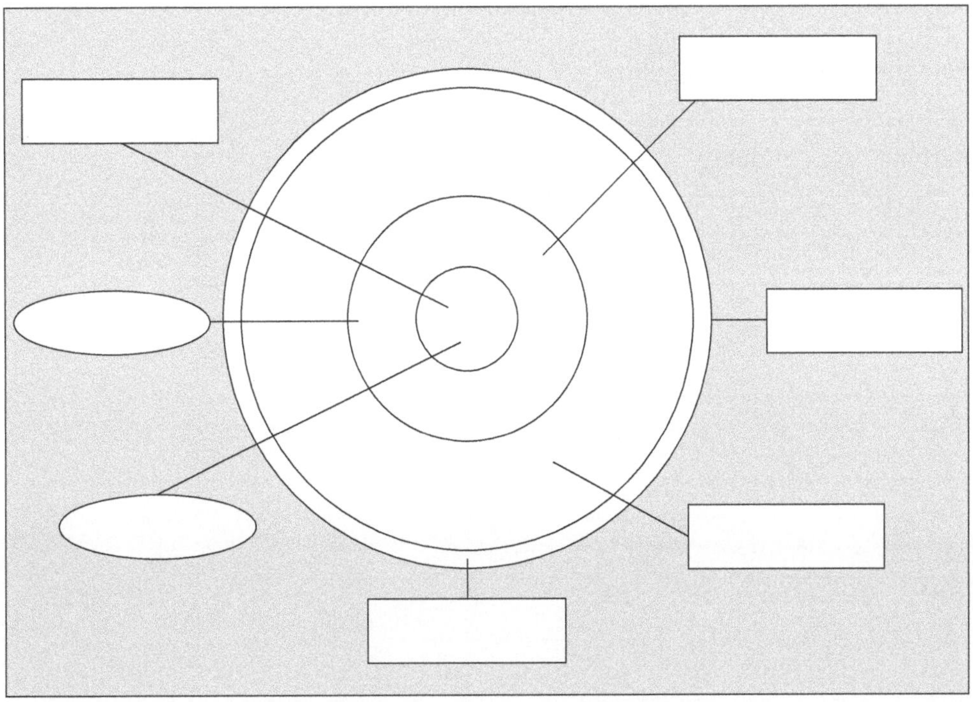

mantle
hard rock
crust
inner core
outer core

b Now colour the inner core yellow, the outer core orange, the mantle brown, and the crust black. Then, add the words **solid** and **liquid** to the diagram.

Now try these...

2 What is the lithosphere composed of? ....

3 **a** What is the radius of the Earth? ....

b How can you use this figure to work out the circumference of the Earth? (Clue: think **pi**)

....

....

c What is the circumference of the Earth? ....

4 There is continental and oceanic crust. What are the differences between them?

....

....

5.2 A cracked Earth?

This is about where the Earth's plates are – and their link with earthquakes and volcanoes.

1 Write a reply to this question. Use these words to help you.

plates	crust	margins	moving	convection

Well, it's because ...

2 Why don't earthquakes and volcanoes happen in some places?

3 Why are the Himalayas increasing in height?

4 Why are volcanoes frequent in Iceland?

5 Why are there no major earthquakes in Britain?

A closer look at plate movements

This is about how the Earth's plates are moving – and how their movements produce earthquakes, volcanoes, and even mountains!

1 These pictures show the different kinds of plate margins.

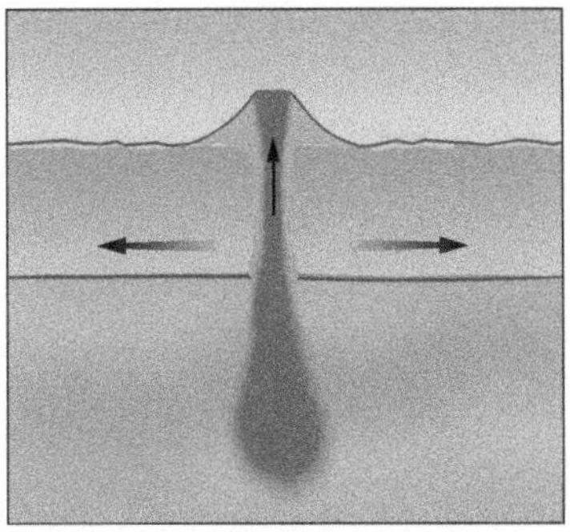
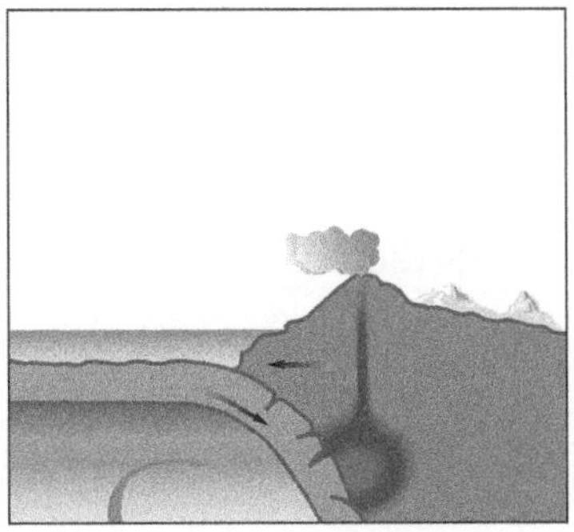
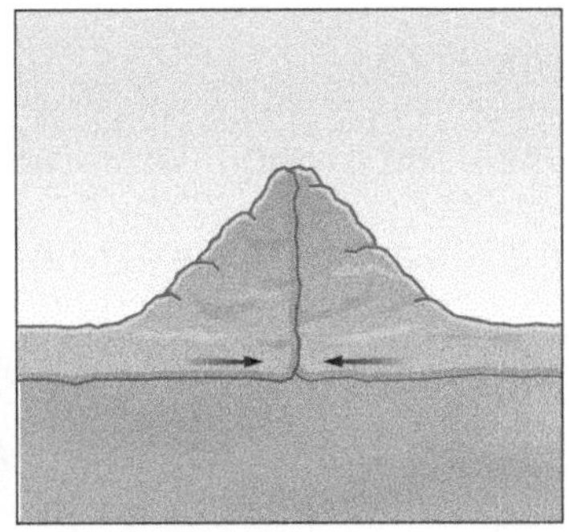
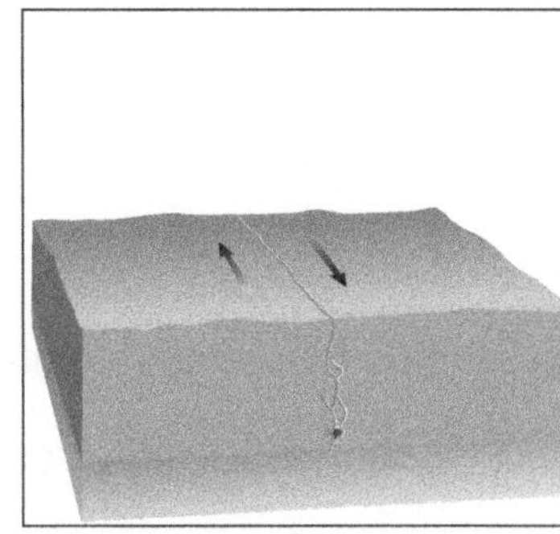

a Label each diagram. Choose from:

- plates are moving apart
- plates are sliding past each other
- one oceanic plate is going under a continental one
- two continental plates pushing into each other

b Draw a red star on the diagrams where you would expect earthquakes to happen.

c Label any volcanoes you would expect.

d Label the fold mountains.

e Label the new crust that's being formed.

2 Which diagram do you think shows a **constructive** (building) margin? Why?

3 Which diagram do you think shows a **destructive** (destroying) margin? Why?

5.4 What is an earthquake?

This is about what earthquakes are, how they are measured, and what damage they do.

Find the correct term to go with each definition.

1 Waves that are made by earthquakes: ....

2 Small earthquakes after the main one: ....

3 The point inside the crust where the earthquake started: ....

4 The point on the surface right above where the earthquake started: ....

5 Amount of energy that an earthquake gives out: ....

6 A machine that measures the magnitude of the earthquake: ....

7 The scale that tells us how strong an earthquake is: ....

8 A tidal wave caused by earthquakes: ....

9 What an earthquake makes the ground do: ....

10 If gas pipes fracture there might be: ....

11 Now find these words in the wordsearch! They could be written horizontally, vertically or diagonally; and forwards or backwards.

r	v	e	e	e	o	u	t	z	w	h	p	u	w	c
w	e	f	m	f	a	c	s	u	s	a	x	f	g	j
w	p	t	o	v	x	p	u	m	y	e	m	j	w	a
v	v	c	e	h	s	c	n	i	d	g	r	q	p	e
r	u	z	b	m	f	f	a	m	s	d	e	i	d	x
s	i	b	a	b	o	r	m	k	m	r	r	u	f	h
q	v	c	t	c	g	m	i	y	p	y	t	d	n	b
v	h	u	h	f	x	f	s	w	u	i	n	b	e	m
o	d	a	y	t	d	q	c	i	n	o	e	s	e	n
e	k	a	h	s	e	v	s	g	e	b	c	k	t	m
h	a	e	t	y	o	r	a	q	p	s	i	m	l	r
t	p	s	e	i	s	m	i	c	p	w	p	k	f	e
s	k	c	o	h	s	r	e	t	f	a	e	c	x	e
d	x	y	f	x	c	x	v	n	d	p	s	n	k	u
b	u	z	p	p	v	h	o	h	c	l	j	y	y	r

5.5 An earthquake in Southwest China

This is about the damage caused in the Sichuan earthquake of 2008.

The earthquake factfile	
date	Monday 12 May 2008
time	2.28 pm
magnitude	7.9 on the Richter scale
epicentre	in mountains in Sichuan Province, Southwest China
damage	over 87 000 dead over 370 000 injured, and at least 5 million left homeless in Sichuan Province
financial cost	nearly £120 billion

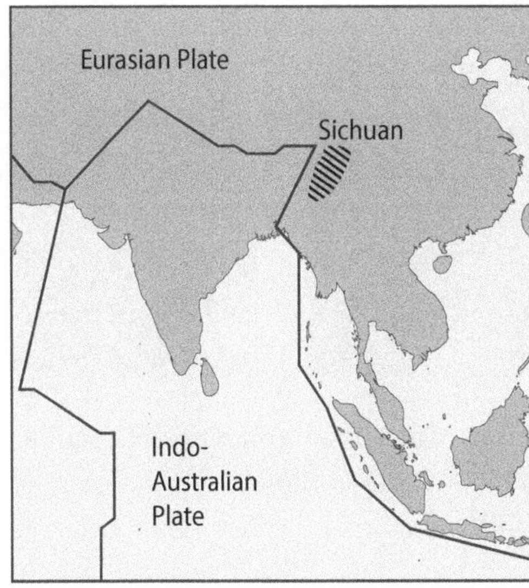

1 Using the information from the map and the factfile, add words to the speech bubble to complete what the two people are saying.

2 Look at the table below. Put a tick in the column to show if what happened could have been stopped or not.

Event	Could not be stopped	Could be stopped
1 The two plates were colliding and causing a lot of strain		
2 On 12 May the strain got too much		
3 At 2.28 in the afternoon the ground shook		
4 Buildings in Beichuan Middle School collapsed		
5 Rock slipped as much as 12 metres along the fault		
6 Parents blamed local officials for erecting poorly built schools		
7 Students and teachers were crushed to death		
8 Over 87 000 people died		

Tsunami!

This is about the damage tsunamis do to the environment, the economy, and people.

A

B

C

D

1 Choose one of the photographs and describe the damage that you can see. Write the letter of the photograph in the space.
Photograph [ ]

....

....

....

....

....

2 Complete the table below to describe the damage caused to the environment, the economy and people.

	Natural (environment)	Economic (business/money)	Social (people)
Photograph A			
Photograph B			
Photograph C			
Photograph D			

5.7 Volcanoes

This is about what volcanoes are, and the damage an eruption can do.

Volcanoes happen on the edges of plates, where one plate goes under another. Sometimes the plate suddenly starts moving and you get a new volcano where there wasn't one before.

This is what happened in Mexico …

Draw a picture for each of the boxes in this cartoon strip.

1 It was 1943. In a flat maize (corn on the cob) field in Mexico…

2 … smoke and ash were seen coming from the ground.

3 The ground cracked and there was a terrible smell like rotten eggs.

4 By the next morning the cone of ash was as tall as a house.

5 Over the next year the cone built up with layers of ash and lava to make a new volcano. It was named after the nearest village – Paricutin.

6 The mountain erupted for 9 years! By the end it had covered 10 square miles in lava and caused over 1000 people to die.

7 The lava and ash caused damage during the eruption but left the soil fertile (good and rich) so now lots of crops can be grown and livestock (farm animals) graze the land.

8 Now Paricutin is nearly 400 metres high.

5.8 The most dangerous volcano in the world?

This is about a potential eruption of Mount Vesuvius.

1 This map shows one plate sinking under a different one, which gives rise to Italy's volcanoes. What are each of the two plates called?

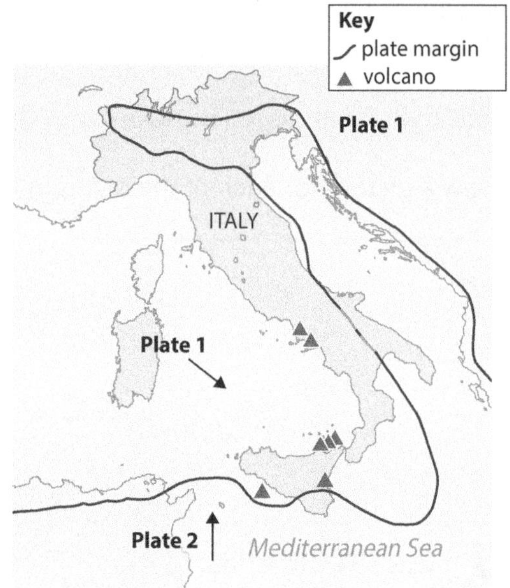

Plate 1: ....

Plate 2: ....

2 Imagine that you are living close to Mount Vesuvius when scientists give warning of an eruption. You are asked to evacuate the area in less than three days. Complete the sentences:

I have to… ....

....

My main concern is… ....

....

When the volcanic eruption is over, I… ....

....

3 Look at this photo of Mount Vesuvius, taken from Naples.
How different do you think it would look after a volcanic eruption?
What would have changed?

....

....

....

....

....

....

....

....

5.9 So why live in a danger zone?

Here we consider issues around living in an earthquake zone.

1 Since 1836 there have been 5 major earthquakes in the San Francisco area, plus 15 smaller but damaging earthquakes. The probability of a magnitude 7 earthquake in the next 30 years is about 75 per cent. You have been offered an exciting long-term job in San Francisco. Give the arguments for and against accepting the job.

For ....

....

Against ....

....

What would **you** decide? Why? ....

....

2 Geothermal energy is renewable and continuous. Ninety per cent of Iceland's energy is produced this way. Why is it more difficult to produce geothermal energy in the UK?

....

....

....

3 In 2022 there was an earthquake in Japan, and the following year there was an earthquake in Türkiye-Syria. Earthquakes occur often in Japan, while they are a lot less common in Türkiye and Syria. Also, Türkiye and Syria are poorer countries than Japan.

The Japan earthquake recorded 7.4 on the Richter scale and there was a loss of life of 3 people. The Türkiye and Syria earthquake recorded 7.8 on the Richter scale and there was a loss of life of over 55,000.

Can you account for the difference in the death toll between the two earthquakes when they were of a similar magnitude?

....

....

....

....

....

....

About Russia

6.1 Meet Russia!

Here you will explain a little more about Russia.

Tip!
Use your imagination, as wel as facts about the country!

1 The overview shown on page 109 in **geog.3** gives six facts about Russia. Imagine that you are a visitor to Russia. Write a postcard home explaining what you expect your visit to Russia to be like.

A postcard from Russia

6.2 What are Russia's main physical features?

Here you will find out more about Russia's 'Mother Volga'.

The Volga River is Europe's longest river, over 2200 miles long, and is seen as the 'national river of Russia'. The river starts north-west of Moscow and flows out in to the Caspian Sea. Eleven out of the twenty largest cities of Russia, including its capital Moscow, are situated in the Volga basin. Over 40% of Russia's population live near it. It is used for transport, hydro-electric power production, and irrigation. Some of the world's largest reservoirs can be found along its banks. It is a major source of Russia's water. The river plays a major part in the day to day life of the Russian people and is known as 'Mother Volga'.

Large quantities of wheat are grown in the fertile river valley, and an important petroleum industry is based in the Volga valley. Other minerals that are found include natural gas, salt, and potash, which is used as a fertiliser. The river's delta and the nearby Caspian Sea offer rich fishing grounds. The Volga also has a number of holiday destinations along its route, and many cruises take place along its course. Nearly the entire length of the River Volga is navigable from about March until mid-December.

1 'Without the River Volga, Russia would not have become such a powerful country'. To what extent do you agree with this statement? Give reasons for your answer.

....

....

....

2 What threats do you think the River Volga could be faced with? Give reasons for your answer.

....

....

....

6.3 Russia's climate zones and biomes

Here you will find out more about Russia's varied climate and vegetation.

1 What is permafrost? What problems may this cause for people?

2 Fill in the gaps in this table. Use the information on pages 112–113 in **geog.3** to help you.

Biome	Main features	Animals
	Covered in ice and snow in winter In summer the surface of the soil thaws	Musk ox, Arctic foxes, polar bears, brown bears, reindeer, ermine
Taiga		Bears, wolves, Siberian tigers
Temperate forest	Mixture of deciduous and coniferous trees	
		Cattle, hamsters, mice
Mountain		Mountain goats, deer, lynx and foxes

3 Which of the biomes would you like to visit? Give reasons for your answer.

6.4 What about the Russian people?

Here you will find out more about Russia's 146 million people.

1 Why is the population denser in the European part of Russia?

2 Why are large areas of Siberia empty of people?

3 Study pages 114 and 115 in **geog.3**. Link the numbers below to a fact about Russia's people. Write the facts in the space next to the number. The first one has been done to help you.

Number	Fact
14%	Percentage of Russia's population who have just enough to survive
75%	
5.5 million	
78%	
185	
146 million	
81%	

6.5 A tour of European Russia

Here you will explore Sochi, a tourist resort on the shores of the Black Sea.

Sochi is the biggest resort town on what's known as the 'Russian Riviera'

Unlike most of Russia, Sochi has a subtropical climate. Its summers are warm or hot, and its winters are mild

You can enjoy beautiful sunsets over the sea, and admire the snow-capped mountains

There are plenty of smart restaurants and outdoor cafes – Sochi has everything you could wish for in a resort

Sochi has the backdrop of the Caucasus Mountains, both pebble and sandy beaches, and the ski resort is 30 miles away

Sochi has more than 200 days of sunshine a year, with an average temperature of above 14 degrees Celsius

Sochi, transformed by the 2014 Winter Olympic Games it hosted, has turned into a year-round resort

Greater Sochi is an incredibly long city – 145 km along the Black Sea Coast

More than three million people visited Sochi during 2014 – the year it hosted the Olympic Games

Sochi: palm-fringed seaside resorts, spectacular ski facilities and a superb conference venue

1 Using the map and an atlas describe the location of Sochi. Remember to use geographical terms in your answer.

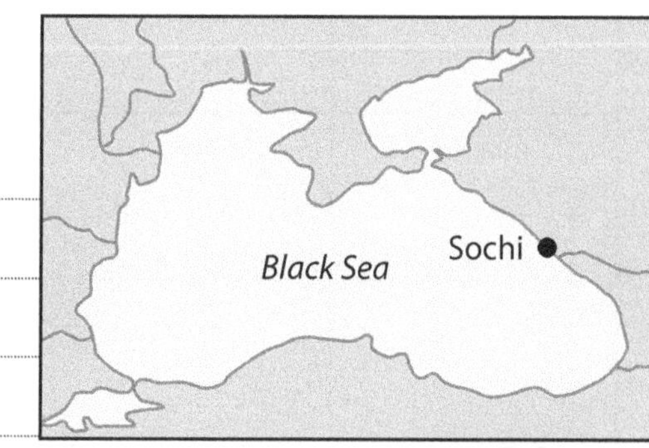

2 The Russian Prime Minister is expected to chair a conference on the development of tourist facilities and mountain skiing infrastructure in the Sochi area. You have been asked to prepare a 100 word news bulletin to help promote the resort. Use the information above to help you write your bulletin.

Tip!
Remember to be persuasive!

Climate change in Yakutia

This is about the features and climate of the Siberian region of Yakutia in Russia.

1 Using an atlas, add the following features to the map:

a the Yakutia district (colour it orange)

b the river Lena (use blue)

c the Arctic Circle (draw a dotted line)

d the city of Yakutsk.

Tip!
Make sure you label all the features you add to the map.

2 Study this climate graph for Yakutsk. A few things are missing!

Using the data from the table, try to complete it by:

a adding the months of the year to the X axis

b adding suitable numbers to each Y axis (average temperature on the left, and precipitation on the right.)

Climate Graph for Yakutia

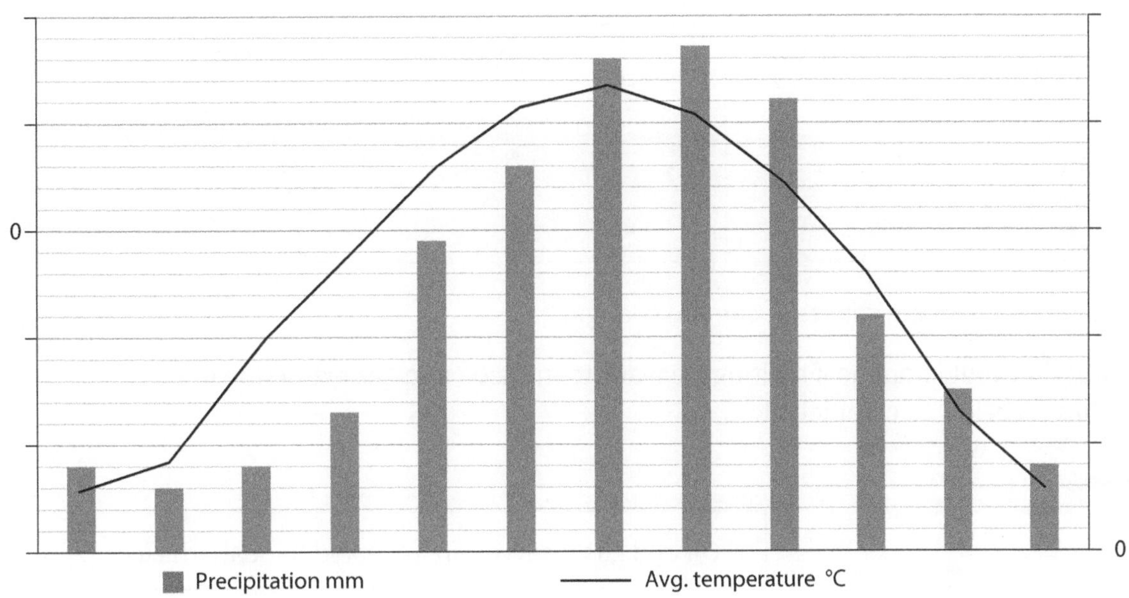

	Jan.	Feb.	Mar.	Apr.	May	June	July	Aug.	Sep.	Oct.	Nov.	Dec.
Avg. Temp °C	–36.3	–32.2	–17.1	–3.6	8	17	20	16	6.4	–6.8	–25.1	–35.9
Precipitation mm	8	6	8	13	29	36	46	47	42	22	15	8

6.7 What is Russia doing in the Arctic Ocean?

This is about the impacts of melting ice in the Arctic.

1 Scientists have noticed that more ocean ice is melting in Arctic summers, and this has been linked to climate change.

How will melting ice affect...

...polar bears?

...ocean ecosystems?

...the Arctic fishing industry?

...Russian oil companies?

...traders in Russia, China, and Europe?

...international tourists?

2 Burning fossil fuels is one of the main causes of climate change.

This flow chart shows what could happen if melting ice leads to more oil and gas extraction in the Arctic. Use the phrases below to complete it.

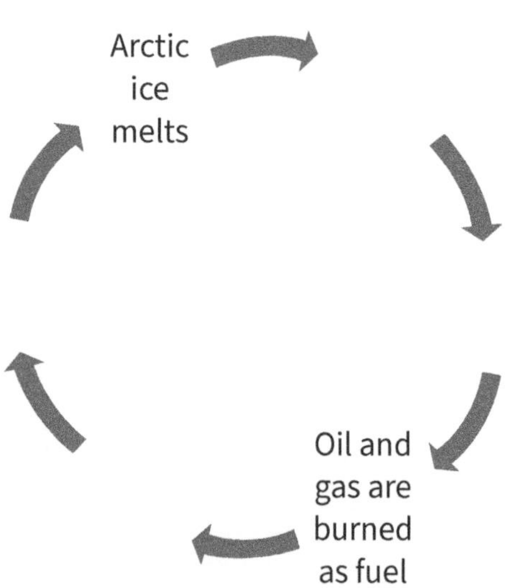

Oil and gas reserves are revealed beneath the ice

Release of greenhouse gases

Extraction of oil and gas

More heat from the sun is trapped, so temperatures increase

The Middle East

7.1 Introducing the Middle East

It's time to find out some facts about the Middle East.

1 Use the map on page 124 in **geog.3** to answer the following

a If you travelled in a straight line from Cyprus to the border between Yemen and Oman, which countries would you pass through?

b Which two countries are found due south of Iraq?

c Which two countries lie south-east of Qatar?

2 Write 'True' or 'False' in the box after each of these sentences.

a The Middle East is where Asia, Europe and Africa meet.

b The United Arab Emirates is made up of eight small states.

c Yemen is an island nation.

d The Middle East's large oil reserves are mainly in Egypt.

e Islam is the dominant religion of the Middle East.

3 A region is an area that shares a feature – such as a distinct climate, or a language, or a mountain range. What do you think defines the Middle East as a region?

4 Research one country in the Middle East. Find out five fun facts and see if a partner can guess which country you have researched.

Fact 1

Fact 2

Fact 3

Fact 4

Fact 5

7.2 The Middle East: physical geography

Here you will find out more about the Red Sea.

1 Read these facts about the Red Sea. Design a fact-filled poster that tells people what the Red Sea is like. You must include a map, and be as creative as you can!

The region surrounding the Red Sea is one of the hottest, driest areas on the planet.

The Red Sea is an extension of the Indian Ocean, located between Africa and Asia.

It is around 190 miles across at its widest point, and about 1,200 miles long.

Intense heat and little precipitation means the water in the Red Sea evaporates quickly.

The Red Sea is home to over two hundred types of coral reefs.

Water in the Red Sea has very clear visibility and warm temperatures throughout the year.

It attracts divers, photographers, marine scientists, and tourists from all over the world.

Cargo vessels, oil tankers, fishing boats, and cruise liners are all found sailing its waters.

Desalination plants along its shores mean that its water can be purified for drinking.

Sea water is also used for industry, such as cement works and oil refineries.

10% of fish species in the Red Sea are not found anywhere else on earth.

Minerals build up in the Red Sea water as it evaporates so the sea is very saline, or salty.

To the north, the Red Sea splits in two, the Gulf of Suez and the Gulf of Aqaba.

The Suez Canal connects the Mediterranean Sea with the Red Sea's Gulf of Suez

The Middle East: climate zones and biomes

Here you will find out more about the Middle East's climate and vegetation.

1 The statements below explain why the Middle East does not get much rain. But they are in the wrong order! Write the missing numbers in the table to show the correct order. Use page 128 in **geog.3** to help you.

Order	Climate fact
1	The land at the Equator gets hot.
	It descends again at the Tropics, around 30° north and south of the Equator.
	The land heats the air.
	The air cools, and its water vapour condenses.
5	Heavy rain falls at the Equator.
	The warm air rises around the Equator.
	The risen air, which has lost all its moisture, gets pushed out of the way.
	The air is dry in the Middle East region.
	As the warm air rises, colder air moves in to take its place.
10	Dry air means little or no rain.

2 Circle the correct biome for the following sentences.

a	Bears and hyena are found here	Desert	Mediterranean areas	Semi-arid areas
b	Plants may have spiky leaves	Desert	Mediterranean areas	Semi-arid areas
c	Scorpions are found here	Desert	Mediterranean areas	Semi-arid areas
d	Sometimes called Steppe	Desert	Mediterranean areas	Semi-arid areas
e	A few trees may be found here	Desert	Mediterranean areas	Semi-arid areas
f	Cypresses are found here	Desert	Mediterranean areas	Semi-arid areas

3 Write fifteen words to explain how climate affects natural vegetation.

....

....

....

The people, and economies

Here you will find out more about the Middle East's population.

1 Write down the correct answer to these questions.

a What is the combined population of Iraq and Iran to the nearest million? ....

b Which country has a population almost exactly six times bigger than Qatar?

....

c Which country has a population eleven times bigger than London? ....

2 What is the biggest ethnic group in the Middle East? List five countries where this ethnic group makes up the majority of the population.

....

....

....

3 Who am I? Identify the two Middle Eastern countries from the facts given below.

I have five letters I am in the south I border the Red Sea I have a population of 30.3 million *Answer*	I am east of the Mediterranean Sea Kurds live in the north My capital is Tehran Most people speak Persian *Answer* ....

4 Middle East Challenge!

Solve the word square below.

arabs	h	n	n	t	e	x	i	z	o	o	c	r	r	a	p
christian	z	a	p	a	p	u	t	l	z	y	z	z	x	t	y
egypt	y	i	m	v	t	y	n	d	h	g	w	j	e	g	a
iraq	u	t	l	v	m	s	g	f	c	u	i	u	n	j	c
islam	r	s	e	u	r	k	i	e	y	b	q	o	o	g	v
jews	j	i	l	a	o	b	i	d	u	j	n	t	m	c	n
kurdistan	e	r	y	r	l	c	b	b	r	a	m	a	l	s	i
kuwait	w	h	j	a	u	i	x	e	b	u	k	j	o	z	t
lebanon	s	c	z	t	d	i	r	e	k	u	k	m	y	i	u
mecca	r	c	t	a	r	h	l	a	w	y	a	t	s	o	r
muslim	c	g	k	q	j	p	f	a	q	n	g	b	r	k	k
oman	m	e	c	c	a	s	i	f	r	x	a	g	z	l	e
qatar	l	c	a	s	w	t	n	z	m	r	v	c	k	h	y
riyadh	o	m	m	l	c	k	l	n	a	f	w	q	l	h	n
turkey	j	g	u	m	u	s	l	i	m	l	v	s	k	a	a

Challenges in the Arabian Peninsula

Here you will find out more about the world's biggest peninsula!

1 Label the map below with three facts about the Arabian Peninsula.

Fact 1

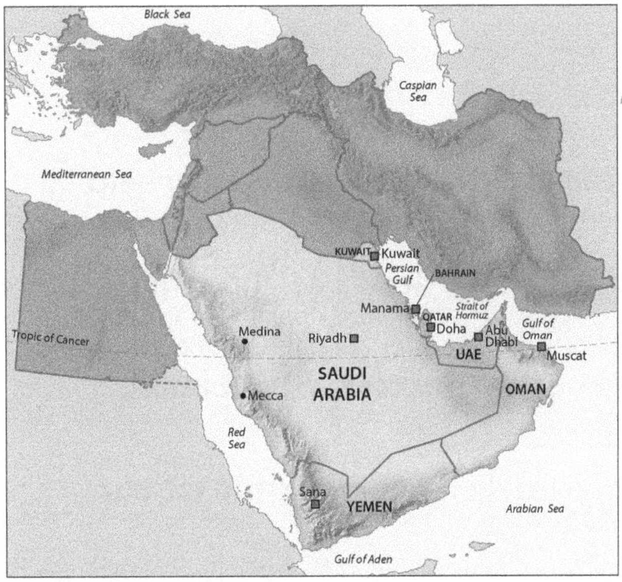

Fact 2

Fact 3

2 The Arabian Peninsula faces a storm of challenges. Look at the challenges listed below, and write at least two ways of tackling each of them.

Water stress: ....

....

Food risk: ....

....

A fall in demand for oil and gas: ....

....

7.6 Conflict in the Middle East

Here you will find out some more about the conflicts found in the Middle East.

1 Conflict always leads to people becoming refugees. What do you think a refugee is?

Over 8.6 million children in **Syria** need **emergency help**.

2.6 million of these children **have fled** to neighbouring countries **in search of safety**.

Over a third of them **do not have access to education.**

An **increasing number** have fallen victim to **child labour.**

2 Read the information above.

Look at some support options given in the table below and rank them 1–5 with number 1 being, in your view, the most urgent.

RANK	Type of Support
	Provide nutritious food for Syrian children
	Offer permanent resettlement to families in other countries
	Provide doctors
	Support education in Syrian refugee camps
	Provide clean water to refugee camps

3 Explain in detail why you have chosen your number 1. Why do you think it will best support Syrian children through this crisis?

Your own notes...

5th edition

geog.3 workbook

This workbook is ideal for homework or independent study. It provides:

- engaging activities
- skills development
- support for every double-page spread in the *geog.3* student book

geog.123 is a comprehensive course that matches the National Curriculum at Key Stage 3.

Did you know?

- The summit of Mount Everest is made of limestone …
- … which formed on the ocean floor!

Did you know?

- We throw out over 500 million small electrical items a year, in the UK.

Ordnance Survey

The world's trusted geospatial partner

kerboodle

Kerboodle provides online Lessons, Resources and Assessment to support teaching and learning in the classroom and at home. Online versions of the student book are also available.

geog.3 workbook – pack of 10
ISBN 978 0 19 848992 4

geog.3 workbook answer book
ISBN 978 0 19 848994 8

Printed on paper produced from sustainable forests.

OXFORD UNIVERSITY PRESS

www.oup.com

How to get in touch:
web www.oxfordsecondary.com
email schools.enquiries.uk@oup.com
tel +44 (0)1536 452620

FSC www.fsc.org MIX Paper | Supporting responsible forestry FSC® C011748

ISBN 978-0-19-848993-1

9 780198 489931